Abandoned Cold Storage Warehouse
Multi-Firefighter Fatality Fire
Worcester, Massachusetts

Investigated by: John R. Anderson

This is Report 134 of the Major Fires Investigation Project conducted by Varley-Campbell and Associates, Inc./TriData Corporation under contract EME-97-CO-0506 to the United States Fire Administration, Federal Emergency Management Agency.

Homeland
Security

Department of Homeland Security
United States Fire Administration
National Fire Data Center

U.S. Fire Administration Fire Investigations Program

The U.S. Fire Administration develops reports on selected major fires throughout the country. The fires usually involve multiple deaths or a large loss of property. But the primary criterion for deciding to do a report is whether it will result in significant "lessons learned." In some cases these lessons bring to light new knowledge about fire--the effect of building construction or contents, human behavior in fire, etc. In other cases, the lessons are not new but are serious enough to highlight once again, with yet another fire tragedy report. In some cases, special reports are developed to discuss events, drills, or new technologies which are of interest to the fire service.

The reports are sent to fire magazines and are distributed at National and Regional fire meetings. The International Association of Fire Chiefs assists the USFA in disseminating the findings throughout the fire service. On a continuing basis the reports are available on request from the USFA; announcements of their availability are published widely in fire journals and newsletters.

This body of work provides detailed information on the nature of the fire problem for policymakers who must decide on allocations of resources between fire and other pressing problems, and within the fire service to improve codes and code enforcement, training, public fire education, building technology, and other related areas.

The Fire Administration, which has no regulatory authority, sends an experienced fire investigator into a community after a major incident only after having conferred with the local fire authorities to insure that the assistance and presence of the USFA would be supportive and would in no way interfere with any review of the incident they are themselves conducting. The intent is not to arrive during the event or even immediately after, but rather after the dust settles, so that a complete and objective review of all the important aspects of the incident can be made. Local authorities review the USFA's report while it is in draft. The USFA investigator or team is available to local authorities should they wish to request technical assistance for their own investigation.

This report and its recommendations were developed by USFA staff and by Varley-Campbell & Associates, Inc. Miami and Chicago, its staff and consultants, who are under contract to assist the Fire Administration in carrying out the Fire Reports Program.

The U.S. Fire Administration greatly appreciates the cooperation received from the Worchester Fire Department.

For additional copies of this report write to the U.S. Fire Administration, 16825 South Seton Avenue, Emmitsburg, Maryland 21727. The report is available on the Administration's Web site at http://www.usfa.dhs.gov/

U.S. Fire Administration

Mission Statement

As an entity of the Department of Homeland Security, the mission of the USFA is to reduce life and economic losses due to fire and related emergencies, through leadership, advocacy, coordination, and support. We serve the Nation independently, in coordination with other Federal agencies, and in partnership with fire protection and emergency service communities. With a commitment to excellence, we provide public education, training, technology, and data initiatives.

TABLE OF CONTENTS

Abandoned Cold Storage Warehouse
Multi-Firefighter Fatality Fire
DECEMBER 3, 1999

Investigated By: John R. Anderson

Local Contracts: Chief Gerard A. Dio
Worcester Fire Department
141 Grove St.
Worcester, MA 01605

OVERVIEW

On Friday, December 3, 1999, at 1813 hours, the Worcester, Massachusetts Fire Department dispatched Box 1438 for 266 Franklin Street, the Worcester Cold Storage and Warehouse Co. A motorist had spotted smoke coming from the roof while driving on an adjacent elevated highway. The original building was constructed in 1906, contained another 43,000 square feet. Both were 6 stories above grade. The building was known to be abandoned for over 10 years. Due to these and other factors, the responding District Chief ordered a second alarm within 4 minutes of the initial dispatch.

The first alarm assignment brought 30 firefighters and officers and 7 pieces of apparatus to the scene. The second provided an additional 12 men and 3 trucks as well as a Deputy Chief. Firefighters encountered a light smoke condition throughout the warehouse, and crews found a large fire in the former office area of the second floor. An aggressive interior attack was started within the second floor and ventilation was conducted on the roof. There were no windows or other openings in the warehousing space above the second floor.

Eleven minutes into the fire, the owner of the abutting Kenmore Diner advised fire operations of two homeless people who might be living in the warehouse. The rescue company, having divided into two crews, started a building search. Some 22 minutes later the rescue crew searching down from the roof became lost in the vast dark spaces of the fifth floor. They were running low on air and called for help. Interior conditions were deteriorating rapidly despite efforts to extinguish the blaze, and visibility was nearly lost on the upper floors. Investigators have placed these two firefighters over 150 feet from the only available exit.

An extensive search was conducted by Worcester Fire crews through the third and fourth alarms. Suppression efforts continued to be ineffective against huge volumes of petroleum based materials, and ultimately two more crews became disoriented on the upper floors and were unable to escape. When the evacuation order was given one hour and forty-five minutes into the event, five firefighters and one officer were missing. None survived.

1

A subsequent exterior attack was set up and lasted for over 20 hours utilizing aerial pieces and deluge guns from Worcester and neighboring departments. Task force groups from across the State of Massachusetts responded to initial suppression and subsequent recovery efforts. During this time, the four upper floors collapsed onto the second which became known as "the deck". Over 6 million gallons of water were used during the suppression efforts.

According to NFPA records, this is the first loss of six firefighters in a structure fire where neither building collapse nor an explosion was a contributing factor to the fatalities.

KEY ISSUES

Issues	Comments
Abandoned building left unprotected and unsecured.	The failure to properly secure and maintain security at this warehouse allowed vagrants to enter, live in, and cause a fire in the building. The lack of detection and suppression systems allowed the fire to grow unrestrained until discovered from the outside.
No barriers to prevent the spread of fire and smoke in a large space.	Despite some floors having over 15,000 square feet of storage space, there were no rated fire walls, functioning fire doors, or even an interior finish that would help limit fire growth and the spread of heat and smoke.
Fire spread via combustible interior finishes.	Being a cold storage warehouse, many walls and ceilings were covered with a combustible insulation material including cork, tar, expanded polystyrene foam, and sprayed-on polyurethane foam.
Delayed fire reporting	The building occupants left the warehouse without notifying authorities, and the fire was reported by passing motorists who observed smoke venting from the roof. The absence of uncovered windows also prevented earlier detection from the exterior.
Access limitations for fire suppression and rescue.	Building construction featured a single staircase from the basement to the roof. This vertical opening was the only way to move through all levels and was congested with men and equipment from the start of operations. The storage areas of the warehouse had no windows. These two factors left firefighters above the first floor without a secondary escape route and prevented ladder and rescue operations through windows.
Unusually long interior travel distances.	Firefighters had to crawl over 200 feet through heavy smoke from the single staircase to conduct a proper search. Most lifelines were only 50 foot and SCBA air was limited to 30 minutes. Searches and rescue operations were ineffective under these circumstances.

THE FIRE DEPARTMENT

The Worcester Fire Department has been a career department since February 2, 1920 when the Board of Engineers voted to staff all fire houses with full-time personnel. In December of 1999, Worcester Fire provided fire protection and first responder level EMS to 172,000 residents in a 37.6 square mile area. The authorized department strength is 485 with ideal manning of 105 firefighters and officers per shift. Minimum manning is 75 firefighters and officers. At the time of this incident there were 469 uniformed personnel. Suppression crews work 42 hours per week on a rotating schedule of two 10 hour days, one day off, two 14 hour nights, and 3 days off. Each group has a Deputy Chief and two District Chiefs who are assigned to the North and South Divisions which divide the downtown area and the remainder of the city in half.

The department operates 15 engine companies and 7 ladder companies out of 12 stations. The ladders consist of 2-75 foot aerial scopes, 4-100 foot tiller aerials, and 1 rear mount, 110 foot tower. The department also staffs a heavy rescue and has available a SCUBA rescue truck and a Hazmat response truck. There are three engines and one scope in reserve.

Minimum staffing is three men personnel truck including one officer. In the absence of an officer, a senior firefighter will assume this function. Trucks in busier areas and the heavy rescue which runs throughout the city have additional personnel.

Upon appointment to the Worcester Fire Department, recruits are subjected to 480 hours of training at the department's fire academy. The training includes Hazmat, CPR, First Responder, rescue as well as the firefighting skills. They must pass a written and practical exam certifying them to the Fire Fighter II level. The Academy and in-service training are coordinated by the Training Division which is headed by a District Chief and is staffed by another three officers and a firefighter.

Fire prevention is headed by a captain who has two lieutenants assigned to fire investigations and two lieutenants assigned to inspections. There are also firefighter level inspectors.

In 1998, the Worcester Fire Department responded to 20,381 emergency calls of which over 40 percent were first responder medical. The department fought 459 structure fires this same year.

PERSONAL PROTECTIVE EQUIPMENT

Each uniformed firefighter is required to have NFPA approved turnout coat, bunker pants, boots, helmet, hood, and gloves. Personal protective equipment is inspected annually by the district chiefs, and each member has a uniform allowance to maintain this clothing as well as station uniforms.

The Worcester Fire Department utilizes Scott SCBA Model 4.5 with 30 minutes tanks. Per their 1974 guideline, the use of SCBA is mandatory for structure first. All SCBA packs include a PASS device. At the time of the fire, 37 packs had integrated PASS devices and were assigned to companies on Rescue 1, Ladder 2, and Engines 1, 2, 3, 4, and 16. All of the fatalities had integrated PASS devices on their SCBA's.

RADIO COMMUNICATIONS

The City of Worcester uses an 800 MHz trunked radio system divided into talk groups. The fire departments uses 5 groups for their operations. The North District Chief switched this incident to FD OP A as is the procedure. Each piece of apparatus has a mobile in the cab and a remote speaker and microphone at the pump panel or aerial turntable. Command vehicles have one portable radio for the Chief Officer and one for the chief's aide. Each engine or aerial has 2 radios, and the Rescue has 7, one for each of the crew, if fully staffed.

As current fireground philosophy dictates, work crews have communications with them as they enter buildings or disappear from the visual sights of Incident Command. Radio communications are used only when direct verbal commands are not possible due to the logistics of the fire scene.

INCIDENT COMMAND SYSTEM

The Worcester Fire Department relies on an established Incident Command System instituted in 1993. All Worcester fires have an Incident Commander and as many of the subordinate command officers as the event dictates. A small event may necessitate only a single Chief Officer to fill all these roles.

The first arriving officer assumes command until relieved by a superior who, in Worcester, is usually the responding District Chief. If the incident escalates, the duty Deputy Chief may then assume command, and the Department Chief may take over if the need arises. Management of an incident may also require the assignment of Sector Commanders to direct specific functions such as ventilation, interior attack, etc.

Staging is also a normal practice and allows for the orderly introduction of crews and equipment. This permits documentation of personnel, their assignment, and their location. It also helps reduce freelancing.

RAPID INTERVENTION TEAM

Since the RIT concept was introduced by the NFPA in 1992, it has slowly worked its way into the fire service. Worcester developed an Operational Guideline in February of 1999, and training had begun within the department.

The standard practice is for the Incident Commander to designate the fourth engine or second ladder of the first alarm assignment as the RIT.

BUILDING CODES

When the original warehouse was constructed in 1906, there was no standard building code in Worcester. Approval and inspections, if any, were done by the fire district in which the property was to be built. The warehouse, as evidenced during this fire, had only one means of egress from all floors above the first. Worcester's "triple deckers", the most common urban residence built in the early 1900's, customarily incorporated two staircases, and one can conclude that awareness of multiple escape routes was present. In this case, the risk was increased due to the lack of windows.

When the addition was erected in 1912, the egress issue was worsened by the substantial increase in square footage on each level and the additional distance that would have to be crossed to get to stairway 3. Current building codes would have required a second staircase.

There are no records of major renovations that would have mandated a life safety upgrade throughout the warehouse. It is unknown if permits were pulled when the office space was added to the second floor of B-building.

Some former employees have stated that access to the southern end of B-building was by the elevators or through the elevators if movement occurred on the same floor through Partition-Y. A few even recalled jumping across the elevator shafts to reach this part of B-building if the elevator cars were committed to another floor. At least one door was installed in Partition-Y on each level after an ammonia leak nearly trapped workers south of this wall. (See Figure on B2.)

BUILDING HISTORY AND CONSTRUCTION

The Worcester Cold Storage and Warehouse building was a six story structure at 266 Franklin Street in the heart of Worcester's former warehousing and cold storage district. In the first half of the 21st century, cold storage was vital to the preservation and delivery of food before refrigerators became commonplace in American kitchens. The location was ideal with rail service provided by the former Boston and Albany Railroad which had a siding against the south end of the warehouse.

Even after the post-WWII decline in railroads, truck traffic was easily accommodated over nearby roads and later on the abutting Interstate 290 which was built in the late 1960's. (See Appendix A for site plan.)

The original warehouse (called "A-building" in previous reports) was constructed in 1906, faced due north onto Franklin Street and bordered Arctic Street to the east. There were six storage levels as well as a basement. The building measured 88 feet by 88 feet and had over 7,000 square feet of floor space on each level. The warehouse had an approximate exterior height of 80 feet.

An addition (called "B-building") was constructed in 1912 against the west wall of A-building and measured 72 feet by 120 feet on the third floor and above. The 72 foot wall faced Franklin Street. The first and second floors were 88 foot and 101 foot deep respectively to accommodate railroad sidings and other structures on the southern on "C" side. Other investigations have referred to the former western exterior wall of A-building as "the fire wall" but there is no indication that this was a planned function. At least one opening was cut through this party wall on each level to access the new addition. B-building provided an additional 7,000 square feet of storage on the third floor and over 8,000 on floors four through six. (See Appendix B for terminology)

The Worcester Cold Storage complex involved additional structures to the south, but these were physically separate buildings and were not involved in this incident. The known openings between the warehouse and the southern structures were for utilities and refrigerants. The only effect was to block aerial access from the south during the fire.

Construction methods appear to be the same in both A and B buildings. Exterior walls were 18 inches thick and consisted of brick and mortar. Interior floors on the first and second levels were poured concrete and were supported by cast iron columns. The concrete was covered with carpet or asbestos tile where appropriate for use. Upper floors were of heavy timber construction with 12 foot long 4 inch by 12 inch wood joists (16 inch o.c.) resting in pockets in the east and west brick exterior walls and attached to 16 inch by 16 inch wood girders on the inside. The girders were on 12 foot centers and rested on 16 inch by 16 inch wood columns which were spaced 12 feet apart in both dimensions. Flooring consisted of two layers of tongue and groove hardwood with some areas having an additional layer of 3/8 inch diamond plate.

Ceilings on individual floors varied from open joists in storage areas to be a suspended ceiling in the office area on the second floor. Photographs taken prior to the fire suggest that some sections also had "glass board" as a finished surface. The exact make up of this material has not been determined. No documentation was made of ceiling heights within the warehouse, but it appears they were approximately 11 foot throughout.

The roof was tar and gravel over a wood deck which covered a 4 foot tall cockloft above the sixth floor ceiling/roof assembly. Roof penetrations included the stairway and elevator shaft on the east end of A-building and a skylight over the elevator shafts on B-building. An illuminated billboard sat on the roof of B-building and received power external to the warehouse structure.

NOTE: For the balance of this report the entire fire building will be referred to as the "warehouse" which consists of "A-building" on the east and "B-building" on the west. The A and B terminology was adopted early on in other investigations and should not be confused with fireground identifications of sides "A, B, C, & D". In a large complex such as this, other terminology could have been created such as "Building 1", "Building Z", etc.

BUILDING USE

Worcester Cold Storage, a business, occupied the warehouse from 1906 until 1983 when it was sold to Chicago Dressed Beef. In 1987, CDB Realty Trust purchased the warehouse. CDB moved its operations to Millbrook Street in 1988 and shut down the refrigeration system in 1989 at which time the building was abandoned.

During its use, various petroleum based insulation materials were incorporated into the building including rigid expanded polystyrene boards and blown on polyurethane foam. These were applied to improve the temperature performance of the buildings Additionally, condensation along the exterior walls lead to the decay of some floor joists. Steel beams or angle brackets were added against the brick walls to pick up the floor load in several places.

Even to long term employees, the building was hard to navigate. The upper four stories were almost identical, and some workers reported getting lost under the dim interior lighting conditions. Condensation would cause ice to form around the ceiling fixtures, and this cone of ice would severely limit the amount of illumination. There was no useful external light then or during the fire.

After it's closing in 1989, the building was illegally entered on many occasions, resulting in vandalism, occupancy by homeless individuals, and a number of small "campfires." At the time the fire occurred, there were no utility services in operation. Significant amounts of garbage and human wastes were scattered around the warehouse. The homeless woman involved in this incident said the interior smelled like a sewer.

VERTICAL PENETRATIONS

There were three stairways in the warehouse. Stairway 1 was in the northwest corner of B-building and went from the first floor (approximate street level) up to the second floor office area. Stairway 2 was located in the southern portion of B-building and went from the first floor to the third. It may have also accessed the basement. Stairway 3 was on the east side of A-building and ran from the basement to the roof. This was the only means of egress from the upper floors and was used heavily during the fire.

Two elevators were adjacent to stairway 3, and two more were adjacent to Stairway 2. At the time of the fire, all had been disabled, and the cars were in the basement. It is unknown if individual access doors were open or closed. The elevator shaft in B-building had a reinforced glass canopy at the roof level.

A 14 inch by 14 inch shaft penetrated the ceiling of the second floor office area and originally housed a 12 inch pipe for the ammonia recovery system. This may have opened through all floors, and the presence of the pipe could not be confirmed.

HORIZONTAL PENETRATIONS

There was one opening on each level through the party wall dividing A-building from B-building. There were numerous doors and windows on the first floor, and several were forced open by firefighters to gain access. All windows on this level were secured with plywood to prevent entry. Windows on the second floor of B-building were limited to the office area in the northwest section and were also covered with plywood. There was a window on each of the second, third, and fourth floors in stairway 3 on the east side of A-building. A window opened into the adjacent elevator shaft on each of these floors also. All were blocked with plywood. (See appendix C for pre-fire photographs)

INTERIOR FINISH

Because the warehouse was used for cold storage, the insides of exterior walls and the roof were heavily insulated. Barriers between office space and freezer space were also heavily insulated. The original material of choice was cork which was impregnated or secured with tar. The thickness has been described from 6 inches to 18 inches depending on the location. Evidence was also found of additional layers of expanded polystyrene sheets and blown on polyurethane. In many places the finished surface was "glass board". A recovered piece of this glass board was ignited by Worcester Fire personnel after this incident. The sample sustained combustion and gave off stringy black smoke not unlike pure styrene.

It has been reported that all the interior partitions were made of corkboard, but it was probably a covering rather than a structural element. The office walls on the second floor were paneling installed over drywall. Many photographs of the cold storage areas taken before the fire show interior surfaces with a clean outer appearance consistent with the glass board. This would have provided a cleanable and wear resistant surface as opposed to bare cork or foam insulation.

INTERIOR LAYOUT

Since the fire did not extend to the basement or first floor, the layout of these spaces is less important. The first floor did, however, provide the access to the rest of the building for fire operations. All space above the first floor was used for cold storage or moving goods with the exception of the second floor office area on the northern half of B-building. (See Appendix D for floor plans)

BUILDING FIRE PROTECTION

Although remnants of a sprinkler system have been referenced, there was no functioning suppression system working at the time of the fire. There was also no detection system in place.

MUNICIPAL WATER SUPPLY

The City of Worcester has an old, but generally reliable water supply system. There is ample supply from reservoirs in neighboring towns, and pressures are strong. In years past, firefighters were known to run attack lines directly off of hydrants by using a gated wye, which is not an accepted practice today.

THE FIRE

The area of origin was discovered to be in an office on the second floor of B-building. (See Appendix E) The cause was determined to be an accidentally overturned candle used to light the room. This happened during a fight between a homeless man who resided in the space and his homeless girlfriend who had recently moved out. She returned to the warehouse in the later afternoon of December 3rd to pick up some belongings and encountered her former boyfriend. A brief shoving match caused the lit candle to fall off a milk crate and into a pile of clothing. Attempts to beat out the flames were unsuccessful, and the fire fed on the combustibles strewn around the room.

The fire started sometime between 1630 and 1745 hours. There was no way to narrow down the ignition time since the couple left after attempting to put out the blaze and did not call authorities. Her pets, a dog and cat, were left behind when they exited the warehouse. One report places a kerosene heater in the area of origin, but fire investigators found no evidence of such a device.

Some physical evidence was found by investigators that supported the statements of the two residents. These included a shadow on the carpet of a pallet which they used as a bed, melted milk crates, and residue consistent with melted candle wax. The remains of the dog and cat were also located. While six counts of manslaughter were brought against each of these individuals, they were eventually dismissed by a Superior Court Judge because Massachusetts law requires "wanton and reckless behavior" on the part of the defendant. The judge felt neither defendant exhibited this behavior, and it is not a crime to "not report a fire." Both of the people involved were believed to have mental deficits.

FIRE SPREAD AND GROWTH

While the initial fuel was ordinary combustibles, it was the unique interior finish that provided the catalyst for this disaster. Investigators determined the walls of the office area were covered with paneling which aided fire growth. While the abandoned warehouse had very little in the way of contents, the fuel load of insulation materials and the glass board was enough to sustain a large inferno. Even without ventilation, there were approximately one million cubic feet of air to contribute oxygen for combustion and harbor smoke.

Once the incipient ignition spread to the tar impregnated cork, polystyrene and polyurethane foams, and the glass board, the fire growth was uninhibited. The fire propagation was assisted by the 225 square foot ventilation shaft created when the rooftop elevator canopy was cleaned out during standard firefighting operations. The fire was discovered because it was already venting through the roof, presumably through broken glass panels on top of the shaft. All windows on the remaining site structures are covered with heavy wire mesh to deflect highway debris and to deter vandalism, but the canopy remained unprotected except for the metal mesh in its design.

Forced entry on the first floor created an ample air supply for combustion since more than one roll up door was opened. An action which is standard firefighting practice. The well developed fire on the second floor had all the elements to free burn, and its size resisted initial suppression effort. Firefighters on the outside of the warehouse stated that it was "raining tar" as unburned globules of tar from the cork insulation were carried up by the smoke column and dropped down after being cooled in the December air. Many pieces of apparatus had to be specially cleaned to remove the streaks of tar, and some Worcester turn out gear still bears scars.

FIRE DEPARTMENT NOTIFICATION AND RESPONSE

Official reports list an off-duty police officer as the person who observed smoke coming from the roof of the warehouse and notified Worcester Fire. Additionally, an off duty Auburn Fire Lieutenant was traveling eastbound on Interstate 290 in the area of College Square when he observed a column of gray/white smoke in Worcester. As he approached, the smoke was clearly coming from the roof of the Worcester Cold Storage Warehouse. He immediately radioed back to Auburn Fire Alarm who notified Worcester. The smoke was not "pushing" but rather rising up in a straight column south of the billboard on the roof, which is consistent with the area of the canopy above the elevator shaft on the southern part of B-building. He also requested that Auburn Fire Alarm notify the Chief (of Auburn) that "this is going to be a multiple alarm fire." Auburn, like the other bordering towns of Leicester, Millbury, and Shrewsbury respond on a mutual aid agreement after a fourth alarm is struck.

Incident times as determined by the Worcester Fire Department's Board of Inquiry are the basis for this report. A transcript of transmissions over "Operations A" channel, the fireground frequency, was one of their objectives in understanding the incident, and the transmission times are logged by the system to an accuracy of 1/100th of a second. (See Appendix F for a summary of these radio broadcasts.)

At 1813 hours on December 3, 1999, Worcester Fire Alarm announced "Striking Box 1438, Franklin and Arctic." Engine 1, Ladder 1, Rescue 1, and Car 3, the North District Chief, were dispatched from Central Street. Engine 6, Engine 12, Engine 13, and Ladder 5 completed the first alarm assignment. In total 7 pieces of apparatus and 30 firefighters responded on the first alarm. Car 3 was at the Greendale Station when dispatched and used I-290 as his response route.

Approaching from the west, Engine 1 reported heavy smoke, but upon arrival saw no indication of fire in the warehouse from the street. This was largely due to the lack of windows, and the smoke column was blocked by the 80 foot walls and the elevated interstate highway. The warehouse revealed virtually no details about its interior. Engine 1 established a water supply with one crew (2 men) while another proceeded up stairway 3 with a Rescue 1 crew to the roof. A Ladder 1 crew and a Ladder 5 crew each climbed their respective aerials to the roof.

The Auburn Fire Lieutenant had exited the highway about one half mile south of the warehouse and, upon parking at the Kenmore Diner and walking to Franklin Street, observed first alarm companies arriving and beginning forcible entry on the A side, first floor. Crews from Ladder 1, Engine 1, Rescue 1, and Engine 13 entered the 1st floor.

Weather conditions reported: 7.5 mph wind from the southwest; temperature of 45 degrees Fahrenheit, and relative humidity of 37 percent.

While responding down I-290 west, Car 3 observed a column of charcoal smoke coming from the roof of the warehouse. Car 3 called for a second alarm and established a staging area under I-290. The second alarm was dispatched at 1818 and brought engines 2 and 16, aerial scope 2, and Car 2, the on-duty Deputy Chief. In all, staffing was increased by 12 firefighters. The ascending Rescue 1 crew completed a quick search of the third floor of A-building. They did not enter B-building.

Engine 13 reported fire in the elevator shaft of B-building at the second and third floors. They were on the first floor of B-building and looked up at the fire through the open elevator doors. A Ladder 1 firefighter reported "a room full of fire" in a freezer room on the second floor and requested a line. Ladder 1's location was the party wall door between A-building and the office area of B-building. The door opened into building B and, once unlatched, swung into the inferno on the second floor pulled in by a strong draft. The Ladder 1 firefighter used his body weight to pull the door closed and contain the blaze while attack lines were being instituted. Smoke conditions in A-building were minimal at this time.

On the roof, a crew from Rescue 1 cleared out the wire reinforced glass canopy covering the elevator shaft in B-building. A 15 foot by 15 foot vent was now in place to release smoke and hot gasses. The smoke had an oily quality. Rescue 1/Portable 3 informed command: "...it's not Worcester Cold Storage. It's the building closest to the highway, we're up at the roof, we have heavy smoke and embers showing."

The observations of this Rescue 1 crew placed a serious fire in the B-building and may have been a motivator for their deep interior searches through the party wall. They would have wanted to reach the D wall near the highway to assess conditions. For some reason, there was confusion about which building was burning. Because they crossed over the party wall, firefighters may have thought they were into a second structure.

At 1824 hours, Fire Alarm advised Car 3 that a citizen reports two people possibly living in the building. The upper Rescue 1 crew (Portables 3 and 5) began a floor by floor search from the roof down while the lower Rescue 1 crew (Portable 1) began a search from the first floor up. The remaining six firefighters on the roof descended stairway 3 to join in suppression efforts. Engine 13 began attacking the fire with a 2-1/2 inch line through the first floor elevator doors. The stream was directed up towards the second floor and was the first water on the fire.

At the second alarm companies staged, Car 2 assumed Incident Command and Car 3 became Interior Command. At about this time, a 2-1/2 inch line was charged on the second floor and directed into B-building through the party wall door. Several water supply problems arose but were rectified in short order, and all pumping engines had ample flows. A second 2-1/2 inch line was established at the first floor elevator.

All first alarm companies with the exception of the Rescue were committed to the interior attack, and a second 2-1/2 inch line was charged on the second floor. Approximately 1000GPM of water was being applied to a well vented fire, and crews were still making little progress trying to move into B-building. Crews forced the loading dock entrance to stairway 2, and the two first floor lines were moved by Engine 13 to the second floor.

The Rescue 1 crew moving up from the ground reported floors 2 and 3 clear. The lieutenant informed Command of heavy smoke on the third floor, but no fire. A photograph shows a closed door in the party wall at the third floor level. (See Figure B2.) The door may have served as a horizontal barrier to the fire and smoke and would have prevented them from seeing any fire on this floor. Heavy smoke may have obscured the very presence of this door.

The Rescue 1 crew climbed up to the fourth floor and reported they could hear the fire but did not see any flames. While on this level, conditions quickly became worse, and the lieutenant and firefighter had difficulty making their way back to stairway 3. Engine 2 was ordered to advance a line up stairway 3 to the third floor.

At 1840 hours, Car 2 ordered a third alarm bringing engines 3 and 7, ladder 2, and 12 additional firefighters. Engine 2 advanced a 2-1/2 inch line well into the third floor of A-building, but encountered no fire. The smoke was heavy, but the line remained uncharged. Five minutes later, Engine 12's Lieutenant ordered all firefighters on the second floor to retreat to staircase 3. The interior conditions in A-building deteriorated rapidly. Some firefighters stated that the atmosphere went from a moderate fire condition to an untenable one within two minutes.

The Rescue 1 crew searching down from the roof began a series of distress calls at 1846 hours: "Rescue 600 to Command. We need help, on the floor, below the top floor of the building, we're lost." This transmission was not answered, so a second was made a minute later: "Rescue to Command. Rescue to Command, we need help on the fourth floor, one floor down, we're running out of air." Without knowing how many floors were in the warehouse, the Rescue crew had no way of specifying their location. They corrected the confusion by repeatedly stating "two floors down from the roof." Engine 1's Captain asked Command to "get everybody out of the second floor, back them out." Firefighters reported that the fire changed from burning vertically to burning horizontally across the ceiling.

Still not answered, Rescue 1/Portable 5 transmitted "Fire Alarm, Fire Alarm! Emergency, emergency! Clear the air, clear the air! Emergency!" Fire Alarm subsequently sent out the alert tones. Other companies continued to use their radios despite the tones and the severity of the messages. With the poor radio performance and a noisy environment, firefighters may not have heard their radios. Command was unsure of the location of the emergency and the identity of those involved. It took three precious minutes for the lost Rescue crew to be identified, and their location was still in doubt.

The Engine 1 Captain repeated evacuation requests for the second floor of A-building as conditions became quickly untenable with smoke and heat banking down from the ceiling. A search for the missing firefighters began using Engine 3, Ladder 1 and Ladder 2. At this point Command also had no way of knowing exactly how many floors were in the warehouse. Searches were being done from the ground up via stairway 3 so continued confusion about the right level persisted. No firefighters interviewed after the fire recalled passing through the opening in the party wall into B-building during one of their searches.

Two serious situations emerged at the same time, deteriorating conditions in A-building's second floor and two lost firefighters. They competed for radio time, and finally Rescue 1/Portable 5 desperately broadcasted "Fire Alarm we have a second emergency here. Get people up on this floor now or we are going to die! We have no air, and we cannot breathe." Although multiple story searches had begun, the Rescue 1 firefighters were too deep in the building to be heard or found in time.

One firefighter from Ladder 2 was ordered to remove plywood covering the stairway 3 windows and separated from his company. The remaining lieutenant and firefighter joined up with the two firefighters from Engine 3 and started a search on the fifth floor. Crews on the fourth floor reported smoke banked down to 1-1/2 feet above the floor and visibility of 5 feet with hand lights.

Engines 2 and 16 used the dry 2-1/2 inch hose as a lifeline to search the third floor, but could only get about fifty feet in (westward) from stairway 3 before running low on air. After exiting the building and replacing their bottles, Engine 16 proceeded to the fourth floor and used a 50 foot rope secured to the rail for a search. Neither method would have allowed them to fully cross the floor of A-building or to enter B-building.

At 1852 hours, Car 2 requested a fourth alarm and notification of the chief. The fourth alarm complement was Engines 8 and 15, Ladder 4 and 9 firefighters. Car 1, the department chief, was notified, and mutual aid fire companies from outside Worcester began to move into city stations. Three minutes later, Car 3 ordered the trapped firefighters to activate their PASS devices.

Rescue 1/Portable 3 replied "They are activated." This was the final audible transmission from the Rescue 1 crew. No firefighter in the building reported hearing a PASS device alarm. The thick insulation probably absorbed sound as well as it held in the heat from the fire. If the party wall door was closed on the fifth floor, searchers would not have been able to hear the PASS alarms. Suppression efforts were ongoing as Engine 15 is ordered to advance another 2-1/2 inch line. Engine 8 and Ladder 4 joined the search on upper floors.

When asked about their location on the fifth floor, Ladder 2's officer was unsure. Two firefighters from Engine 3 and a firefighter and lieutenant from ladder 2 who were working together in a search appeared to be lost. The lieutenant requested assistance in locating stairway 3 where conditions were also degrading as it began to serve as a second chimney for the fire. None of the four men transmitted a message about hearing a PASS alarm despite their proximity to the missing rescue crew.

While firefighters on the second floor attacking through A-building were forced back by the smoke and heat, the crews in stairway 2 may have been getting some relief from the 225 square feet of the elevator shaft next to their position since this served as a powerful vent during the fire. Five 2-1/2 inch lines were now pouring on the second floor fire, but officers reported no progress in knocking it down. The water flow could have been as much as 1250 gpm into a space with 66,000 cubic feet of volume. Firefighters encountered numerous wires hanging from the ceiling and often got entangled. These may have been communication or electrical lines put in during the office renovation. Engine 13's Captain observed fire dropping down around the structural columns of the second floor of B-building, that may have been burning foam insulation.

Car 1 assumed Incident Command at 1913 hours, and Car 2 became Operations. Car 1 ordered a head count. Ladder 4 tied off their 50 foot lifelines to the dry 2-1/2 inch line on the third floor and searched up to the party wall which they thought was the D side. They heard the sounds of fire near the wall.

At 1915 hours, Car 2 ordered all companies to cease operations to lower the noise levels and help the fifth floor crews identify their exit. Later that minute, ladder 2's Lieutenant made his final transmission: "Ladder 2 to Command we're done…." No indication of danger or entrapment was given up to this point.

The Lieutenant from Engine 3 proceeded to the fifth floor. Once in the stairway vestibule, he listened for any sounds from the firefighters, but heard none. He yelled to help lead the firefighters out. None came. A single Ladder 5 firefighter, secured to a lifeline, went down stairway 3 from the roof to the sixth floor and banged his ax against the steel railing. He fought through the heat to the fifth and fourth levels and repeated the banging. There was no response. He met an officer and firefighter from Engine 3 in the stairway. The Engine 3 Lieutenant had yelled as loud as he could for more than five minutes without any answer and exited with little air left. The ladder man returned to the roof where it was difficult to find the aerial in the heavy smoke.

At 1926 hours, Car 1 radioed: "Give me a fifth alarm on this right now, a fifth alarm." The assignment brought Engines 5 and 10 with firefighters and Car 4, the south District Chief with an aide. Car 1 designated these firefighters as the RIT and ordered them to the B side of the warehouse. Within five minutes, Millbury Fire department arrived with their thermal imaging camera. The ladder 5 crew on the roof observed the tar bubbling on the roof of B-building. They descended to the ground over their aerial.

The Millbury and Worcester personnel climbed stairway 3 to the fourth level but were forced back by heat. They knew they could not reach the fifth floor. Flames had involved A-building on the second floor and were making stairway 3 nearly impenetrable. The thermal imaging camera stopped working, and the Millbury Assistant Chief who was using it reported that the entire screen went black. Thermal overload was suspected in the high heat. The Assistant Chief exited the building and attempted to restart the camera without success.

Exterior units reported a six foot crack in the brick wall at the top of the D/A corner, and the warehouse's structural integrity was in question. Firefighters recalled hearing several "booming" noises that shook the building. At 1951 hours, fire vented through the roof of B-building. Heavy fire was reported with flames 30 to 40 feet above the roof.

At 1958 hours, Car 3 ordered the evacuation of the warehouse. Apparatus air horns were blown and all crews left the warehouse. A final entry was attempted by the South Division Chief and three firefighters up stairway 3 to the third level. Extreme heat halted at that point in operations, their advance some ten yards into the third floor preventing further rescue attempts. And six firefighters were unaccounted for and presumed dead.

Fearing a building a collapse, fire apparatus was repositioned and firefighters dug in for an extensive suppression effort. The exterior attack continued for the next 20 hours with nearly a dozen master streams wetting the inferno at any given time. The fire was contained to the warehouse with no extension, and the flames were knocked down enough by the next afternoon to permit a crane into the scene. An all night effort removed enough of the A/B corner walls to allow the start of recovery efforts. Aerial streams were kept in place for several days to quell occasional flare-ups or exposed pockets of fire.

During the lengthy exterior attack, the warehouse's interior collapsed onto the concrete floor of the second story which became known as the deck. The heavy columns and concrete floors of the first and basement levels were strong enough to resist collapse. All subsequent recovery efforts took place on the deck as Worcester firefighters removed debris by hand to recover their lost colleagues.

FIREFIGHTER FATALITIES

The first body was recovered on Sunday morning in A-building, but it would take six and one half days to find the last. The recovery effort involved the systematic removal of walls and debris, careful sorting of debris, and regular suppression of flare-ups. While most of the recovery effort was performed by Worcester crews, departments and firefighters from across New England came to the scene to assist or cover Worcester fire stations.

Officials hoped the recoveries would happen quickly since the first firefighter was found so early on. As it turned out, only one other firefighter was found in A-building. (See Appendix G) Fearing that the inferno had incinerated most of the deceased, secondary sifting operations were set up around the site under the direction of the Medical Examiner's Office. These washing and sifting areas ran twenty-fours hours a day and were manned by firefighters across New England. Fortunately, the four additional firefighters were found in B-building as that area became accessible.

At 2227 hours on December 11, Fire Alarm transmitted the recall of Box 1438.

LESSONS LEARNED

1. **Abandoned buildings remain a serious threat to the fire service and a danger to the communities in which they stand.**

   Fire departments have long recognized the danger of abandoned buildings in their communities, and fires in these structures have to be approached with a certain amount of caution and restraint. If questionable structural integrity, unknown hazardous materials, unusual dangers to firefighters, or other extreme risks exist, the buildings should not be entered. It is paramount that the fire service apply tactical risk assessment in its daily operations.

   Because of the building design, the fire's magnitude and location could not be ascertained from the exterior, and the Incident Commander had to assess the risks of sending in teams to evaluate the fire and sending in firefighters for suppression. Initial interior reports did not indicate a serious threat to personnel, and operations were conducted accordingly.

To assist arriving crews, a placard system should be instituted which clearly defines the risks at an abandoned building. Subsequent to the fire, Worcester Fire put such a system in place. The process has an added benefit of placing firefighters and/or inspectors on locations which might be at risk and where prefire planning should be initiated.

Risks are not limited to the fire service. Homeless people and drug addicts have been known to inhabit such buildings out of necessity. Ordinary citizens can be impacted by increased crime, and these properties can become a very dangerous playground for inquisitive children. Efforts should be made to renovate or demolish such places even if public funding is not required.

2. **Firefighters must make a concerted effort to know the buildings in their response districts.**

Commercial buildings, by their very nature, pose additional dangers to firefighters, and their familiarity with any given fire building will help to lower these dangers. Company tours are an excellent way to accomplish this goal, and can serve to strengthen the bonds between firefighters and business owners. Such efforts must be conducted with sensitivity, and observed conditions or problems within a business should be conveyed in a helpful rather than confrontational manner.

3. **Fire prevention efforts should be maximized in abandoned and temporarily vacated building to avoid fires in the first place.**

Even temporarily vacated properties can be at risk if utilities like water for a sprinkler system or electricity for an alarm system are disconnected. Although service cessation often occurs when properties are the subject of financial problems it may also take place at the end of a lease or during the sale or renovation of a commercial building. Every effort should be made to forward change of occupancy or use information to first response stations.

4. **Fire departments should continue to grown their file information on buildings in their communities.**

Through the use of mobile computer systems, much information can be forwarded to responding companies and Incident Command during an emergency. Data could include floor plans, occupancies, hazardous materials, water supplies, special hazards, and much more. A system of this type would certainly not be limited to abandoned buildings, but it could be invaluable at such a scene since the probability of an owner showing up is unlikely.

Although this is laborious process, it may also be a valid use of on duty personnel who can gather information during regular shift time and either forward it to fire prevention or enter it themselves on provided computer terminals. Data could be gathered during in-service inspections and tours.

5. **Delayed reporting allowed the fire growth to exceed the capabilities of aggressive interior attack suppression.**

The exact time of ignition remains an unknown, but it has been established that the fire was burning for a minimum of 25 minutes before smoke was observed venting from the roof. It could have been burning for over an hour and a half. The huge volume of air in the warehouse could supports a large fire without any additional air from the outside.

Because flames weren't visible from the exterior, passers-by did not recognize the presence of the fire, and it wasn't discovered until smoke vented from the roof. Even that was apparently not enough to motivate the hundred of average citizens driving on I-290 that evening to call 9-1-1.

The trained eyes of public safety professionals were needed to separate this from "the ordinary" and then react appropriately. By this time, however, most of the second floor of B-building was burning, and few barriers were present to prevent further growth.

The initial report from Ladder 1 on the second floor describes a "room full of fire" in B-building beyond the door in the party wall. This location is some 30 feet from the room or origin, so a one room fire had enough time to engulf the entire floor. A sustained flow of 1000 GPM for 20 minutes had virtually no effect on the fire, and conditions deteriorated around attack crews. (See Appendix E – Room of Origin.)

6. **Combustible interior finishes contributed to the rapid fire spread.**

The concept of having 18 inches of combustible materials on the inside of all exterior walls of a building is almost unthinkable to firefighters. The original cork insulation which appears to have been attached with a tar-like substance provided a large volume of fuel, and additional layers of polystyrene and polyurethane with there ferocious burn characteristics gave this fire enormous intensity.

The area of origin was office space converted from a cold storage area. Under its original design and intent, insulation would only have been placed on exterior walls since the third floor was also cooled. Large amounts of insulation were put into place during the transition and would have included heavy insulation above the suspended ceiling on the underside of the third floor deck. An easily applied insulation would have been sprayed-on polyurethane foam which would have adhered to the wood joists and girders. Once the ceiling tiles were in place, it would not be noticed. The southern wall of the office space would have also required substantial insulation to keep out the cold and to retain the forced hot water heat from the radiators.

The fire fed on ordinary combustibles during its initial growth, but once the ceiling tiles were breached, flame contacted combustible wire insulation and ceiling insulation. The stubborn flames observed by fire crews and the smoke conditions described on upper floors are consistent with the sustained burning of petroleum based products including rigid polystyrene, polyurethane, tar, and glass board.

Proper permitting and on going inspections for construction changes within business occupancies can help reduce non-complaint interior finishes.

7. **The fire service should initiate life safety activities early on at a fire scene.**

The concept of a Rapid Intervention Team was known to the Worcester Fire Department and was being implemented before the Worcester Cold Storage Fire, but it was not put into place until the 5th alarm on December 3rd. Firefighters had entered an unknown structure over one hour before the team was assigned. It is now standard procedure in Worcester to assign a RIT at the onset of each structure fire attack.

The first radio transmission by the Safety Officer was 10 minutes after the RIT was assigned. For control and monitoring of personnel, structural integrity, and other safety concerns, this position should also be filled early on. In an ideal fire scene, the Safety Officer and RIT would be in place before the first firefighters enter the building. Command should strive to have these jobs filled as early as possible even if doing so escalates the event to a higher alarm level to provide sufficient personnel.

A system of personnel accountability should be in place. Someone should be tracking who enters the building, the time of entry, and time of exit. Firefighters who are nearing expected times of air exhaustion could then be contacted to ascertain their safety. The establishment of a Safety Officer at the onset of an event can work towards the goal of accountability. The Safety Officer need not be a department officer but could be a chief's aide or available firefighter familiar with the duties and responsibilities of the assignment.

8. **Large buildings such as warehouses and highrise merit unique search techniques and tools.**

While the standard air bottle for SCBA has a 30 minute capacity, it might be necessary to have available 60 minute bottles for extended search situations and/ or RIT use. Some fire departments have obtained 60 minute systems for use in confined space rescues or other unusually long events. The 30 minute system has remained the norm in recent years as the necessity of Rehab time has gained prominence, and it would not be advisable to use longer air supplies on a regular basis.

In high rise incidents, it is common practice to carry in extra SCBA bottles. The same can be done in large space searches. Development of equipment and techniques to change bottles in a hot environment would give extra range to rescuers, and it could prolong their survival should their own rescue be required.

Long lifelines should be maintained for entry crews in these types of structures as well as marking devises for the interior. These devices include luminescent stickers to show direction, labels to signify searched areas, and other commercially available products. Their effectiveness, however, depends on their use. And the fire service should incorporate these procedures into more common firegrounds, such as single family houses. The time to try out a new technique is not during a major fire scene.

For searches involving extended distances, it might be helpful to position secondary search teams part way into a search area. They can wait in reserve in case they are needed, and they can serve as a rescue team for civilians or firefighters.

Finally, all firefighters who enter a structure must be wearing an SCBA. Worcester Fire has such a policy. Although the facemask and air may not be needed, it must be available. This includes chief officers, aides, and ladder personnel. Even firefighters who are outside structure like apparatus drivers should have SCBA protection available in case of wind shifts or air born particles and debris. With the preponderance of hazardous materials in businesses and residences, SCBA's use is an essential.

9. **Techniques must be improved to better track the movements of firefighters within a structure.**

Under current technology limitations, Incident Command is essentially limited voice communication/radio to track the movements of firefighters once they enter a building and disappear from sight. IC normally knows where a crew entered and possibly what their destination is, but without good radio reports, the exact movements and locations of crews are uncertain at best.

Rescue 1's crew and Engine 3's Lieutenant both had difficulty communicating their positions which complicated and delayed rescue attempts. Crews continued to search multiple floors in the warehouse because of this uncertainty tying up precious personnel resources and adding more congestion to Stairway 3.

Despite all lost firefighters wearing integral PASS alarms on their SCBA's, no surviving firefighters recalled hearing them at any time. The building insulation may have absorbed much of their sound, and the ever present background noise of the fire scene itself may have obscured the rest.

10. Radio channels are often overloaded at multiple alarm fires, and alternatives must be explored.

The 800 Mhz trunked radio system used by the Worcester Fire Department had several major failures during this event. Mechanical failure of individual units occurred when the "emergency alert" button on the hand microphone shorted out on contact with water. Fire Alarm repeatedly ordered individual radio operators to shut down, and this took precious air time during an escalating multiple alarm event. In some cases the microphones were detached in the field at which time they functioned normally. Microphones without the alert button were placed on all radios after the conclusion of this fire. During interior operations, there were 1,000 "push-to-talks" registered for the Operations A talk group, the assigned fireground channel.

Like many progressive fire departments, Worcester has taken steps to insure that all crews entering a fire building have radio communications. A typical piece of apparatus carries one portable for the officer and one for a second firefighting crew. All members of the Rescue Company carry portables. Having multiple radios is good for safety, but their use requires significant training and discipline. It is all too easy to clog up the air with nonessential transmissions.

In some events it may even be necessary to use more than one radio and frequency to properly manage the incident. This would require someone to assist the Incident Commander and keep communications in order. If nothing else, a fireground frequency must be adopted by Command and all working units. One possible way to limit talk time would be to have a staging officer communicate with, and pass along assignments to incoming companies on a frequency other than those used for dispatch and fireground command. Once an assignment was initiated, the company would switch over to the fire- ground channel.

Departments must also choose their radio equipment carefully. The band used must be the best for the standard physical environment in which operations are conducted. Urban departments working inside cement buildings have requirements that contrast greatly with a rural department operating over long geographical distances. If transmission quality continues to suffer, the use of mobile repeaters or other devices might need to be explored.

11. The use of Thermal Imaging Cameras should be further developed.

The Thermal Imaging Camera has become a useful rescue and investigative tool for the fire service over the past six years. Although early models had some operational problems, the latest versions are reliable and offer more options such as transmission capabilities. It is a device that belongs in every fire department, but its high cost has prevented the purchase by many agencies. Sales volume will hopefully bring down the price of this beneficial tool.

The camera used at the Worcester fire failed to operate properly, and the manufacturer attributed the problem to thermal overload. This was an early model, and the rescue crew using it was nearly prevented from entering the warehouse by the high heat. Their attempt to enter was one of the last, and no other crews made significant interior progress.

Under this high heat, the effectiveness of the device is questionable. Thermal imaging devices work well in cooler environments where the body temperature of a victim is higher than the surrounding air or a hot spot within a wall is warmer than the abutting construction. At high heat levels, these cameras will often "white out" because everything in its view is hot enough to affect the imager. If a victim was down in elevated heat, he would absorb the thermal energy of his environment. The turnout gear, for instance, would get hotter and the camera would not be able to differentiate between it and its surrounds. The survivability of a person in high heat for an extended time is negligible.

APPENDICES

APPENDIX A

Site Plan

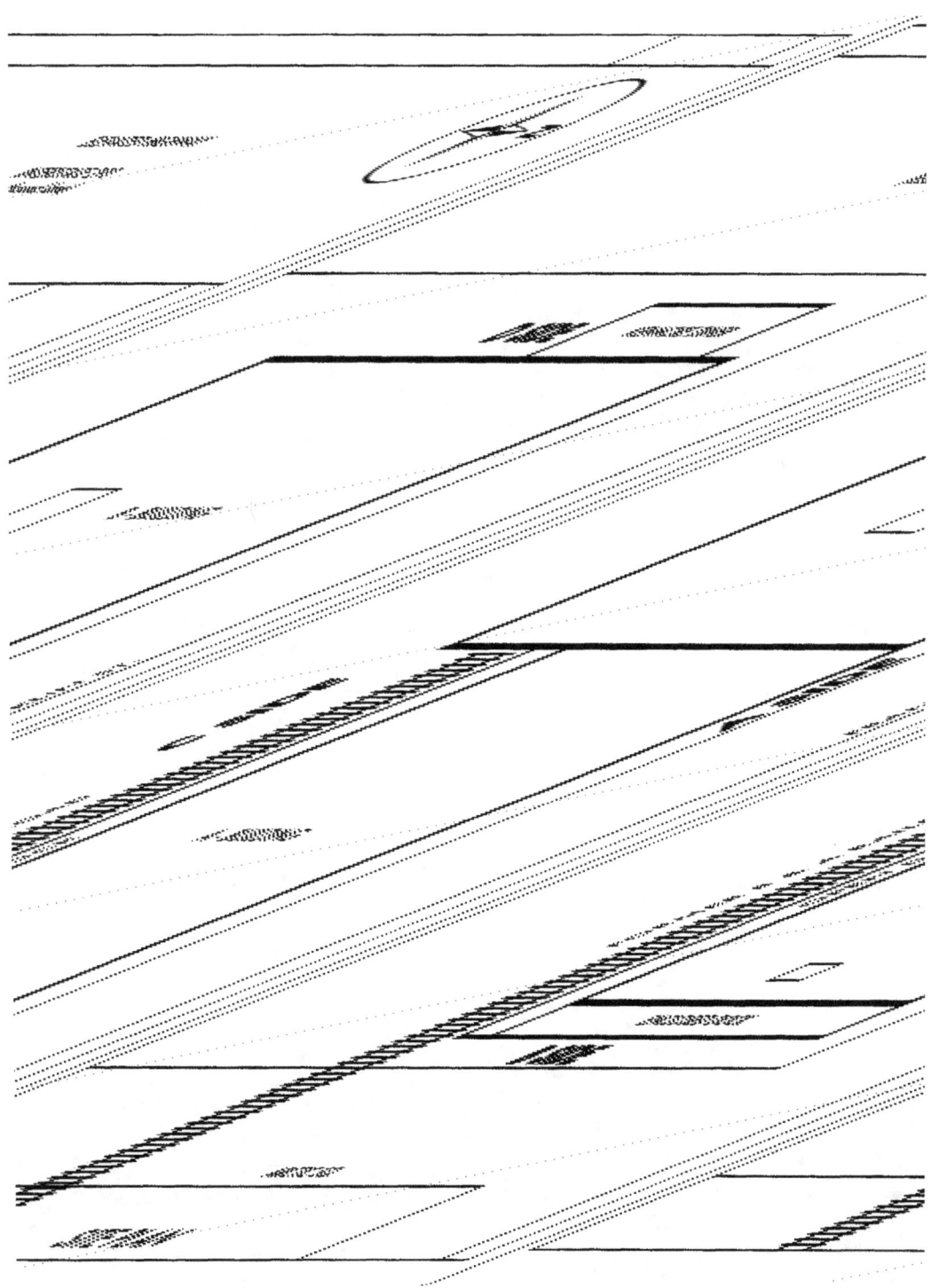

APPENDIX B

Terminology

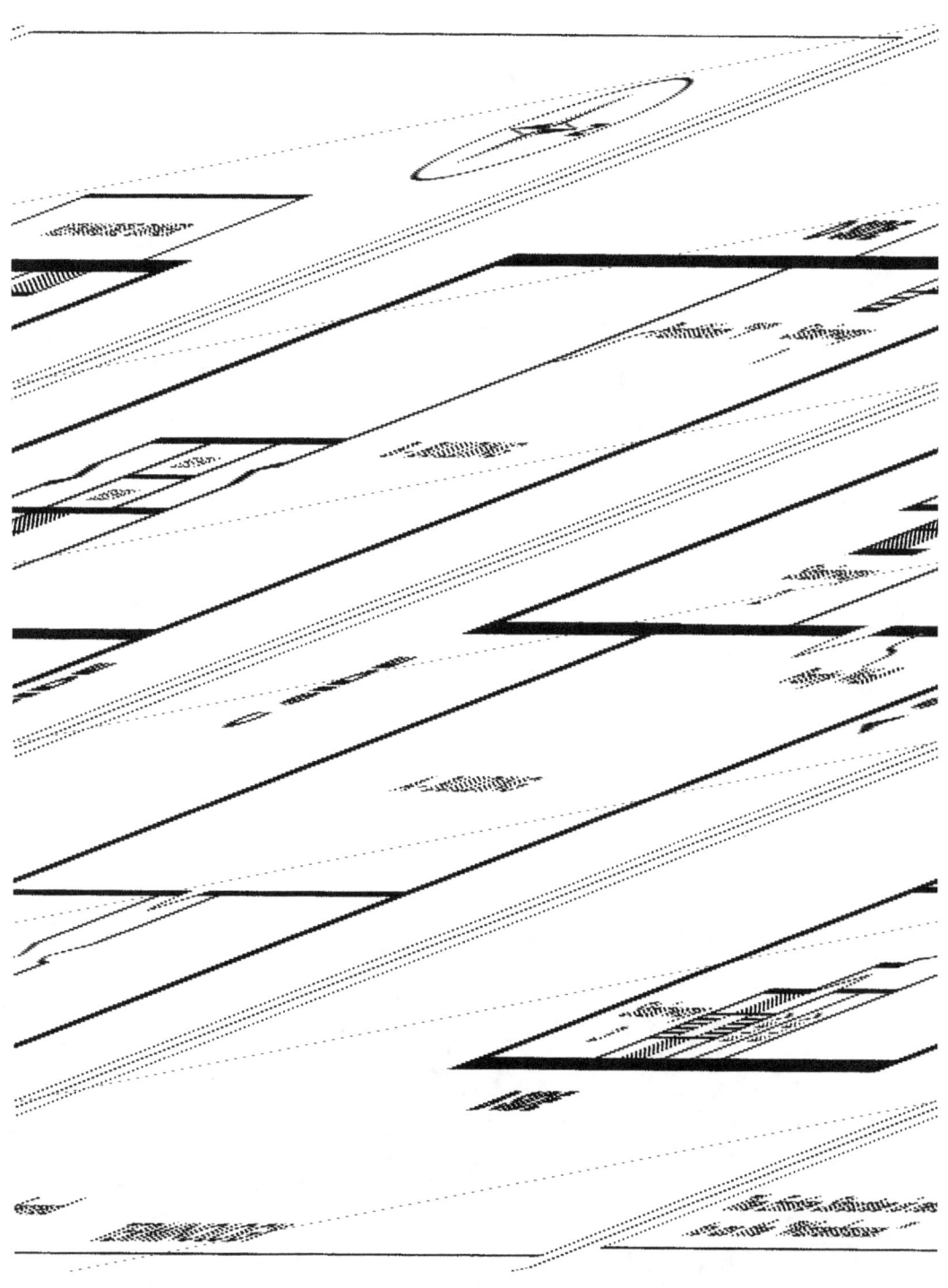

APPENDIX C

Building Photos

B-building A-side (1998)

A-building A-side (1998)

A-building B-side (1998)

B-building B/C Corner (1998)

B-building D-side (1998)

Warehouse D/A Corner

APPENDIX D

Floor Plans

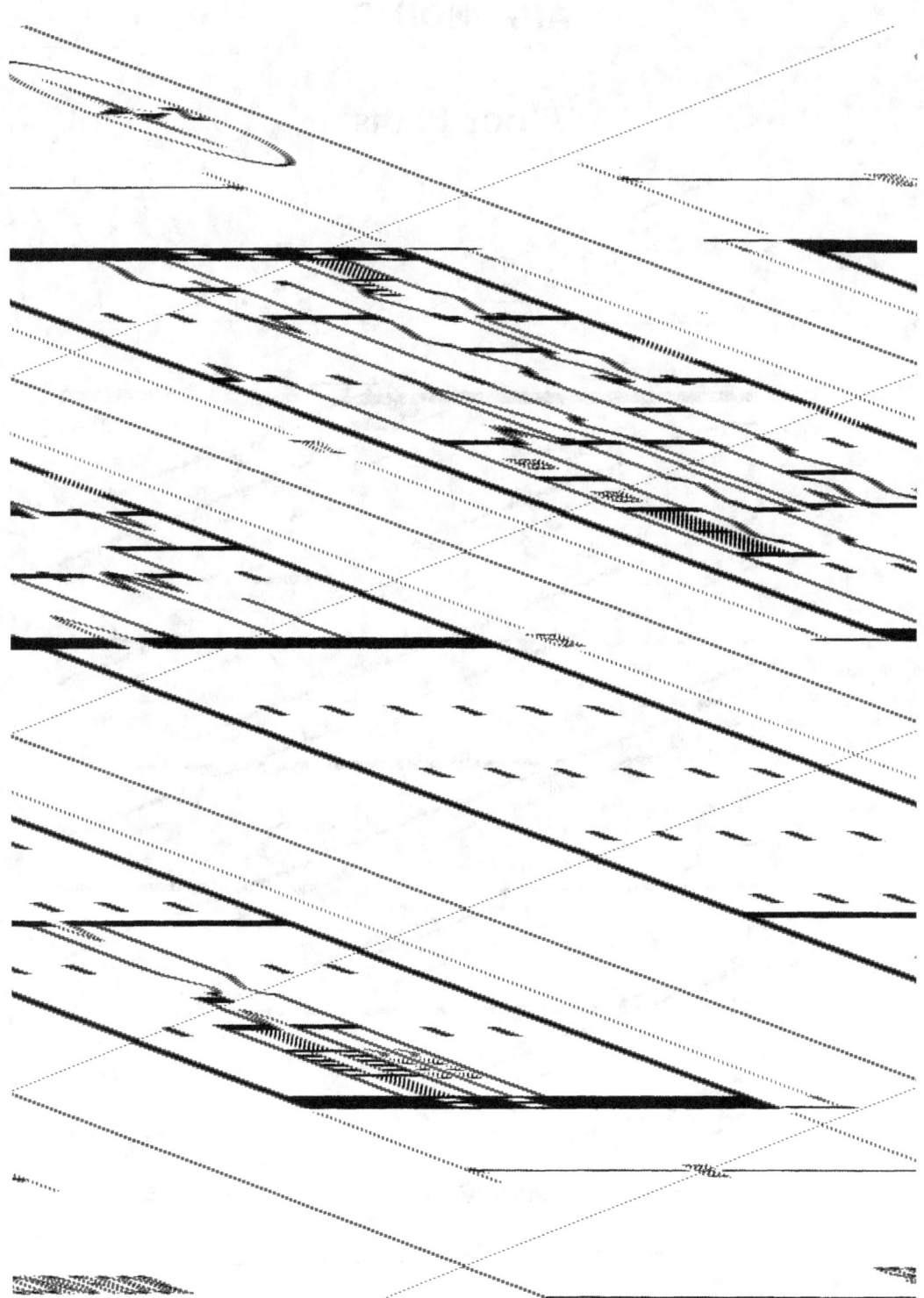

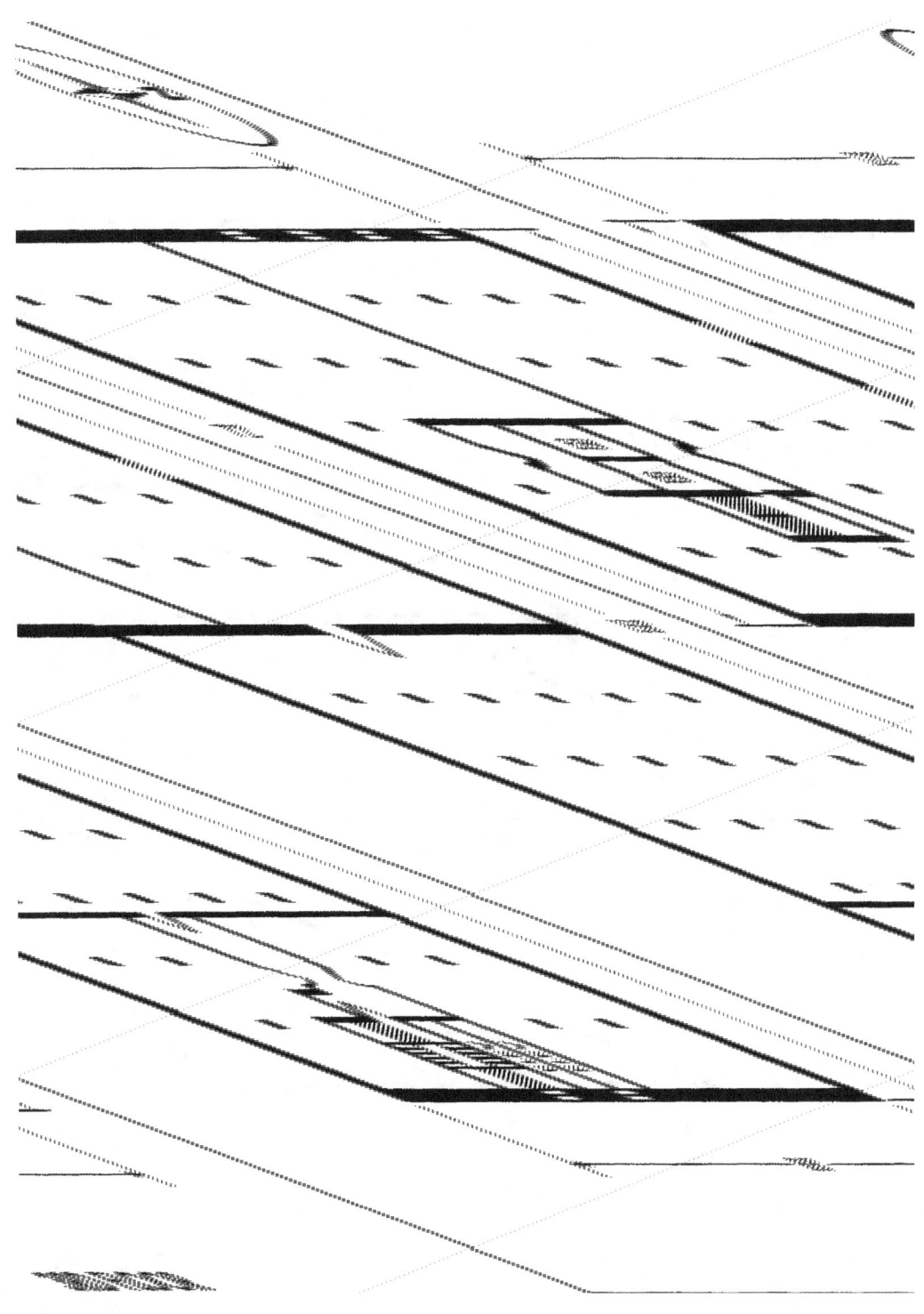

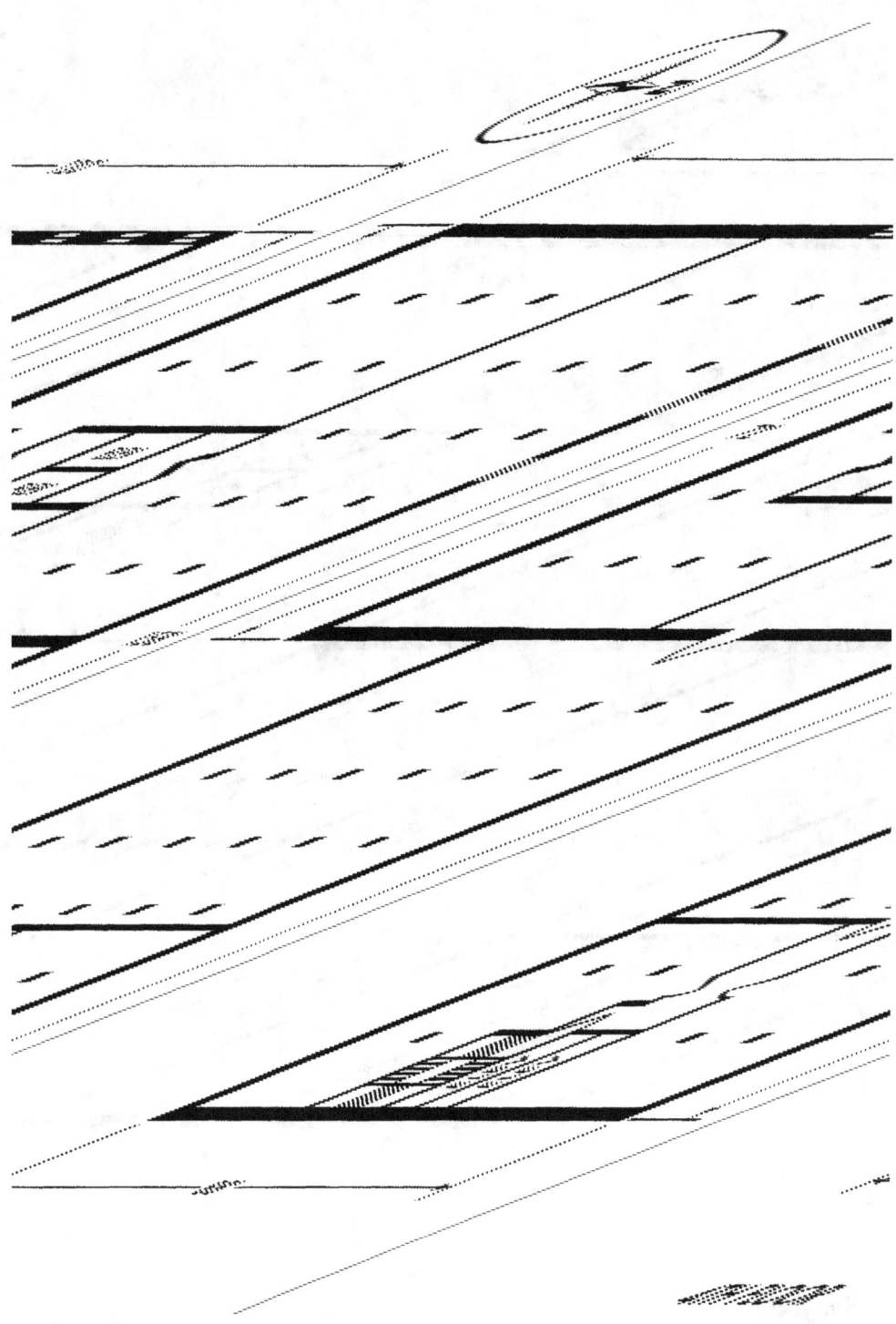

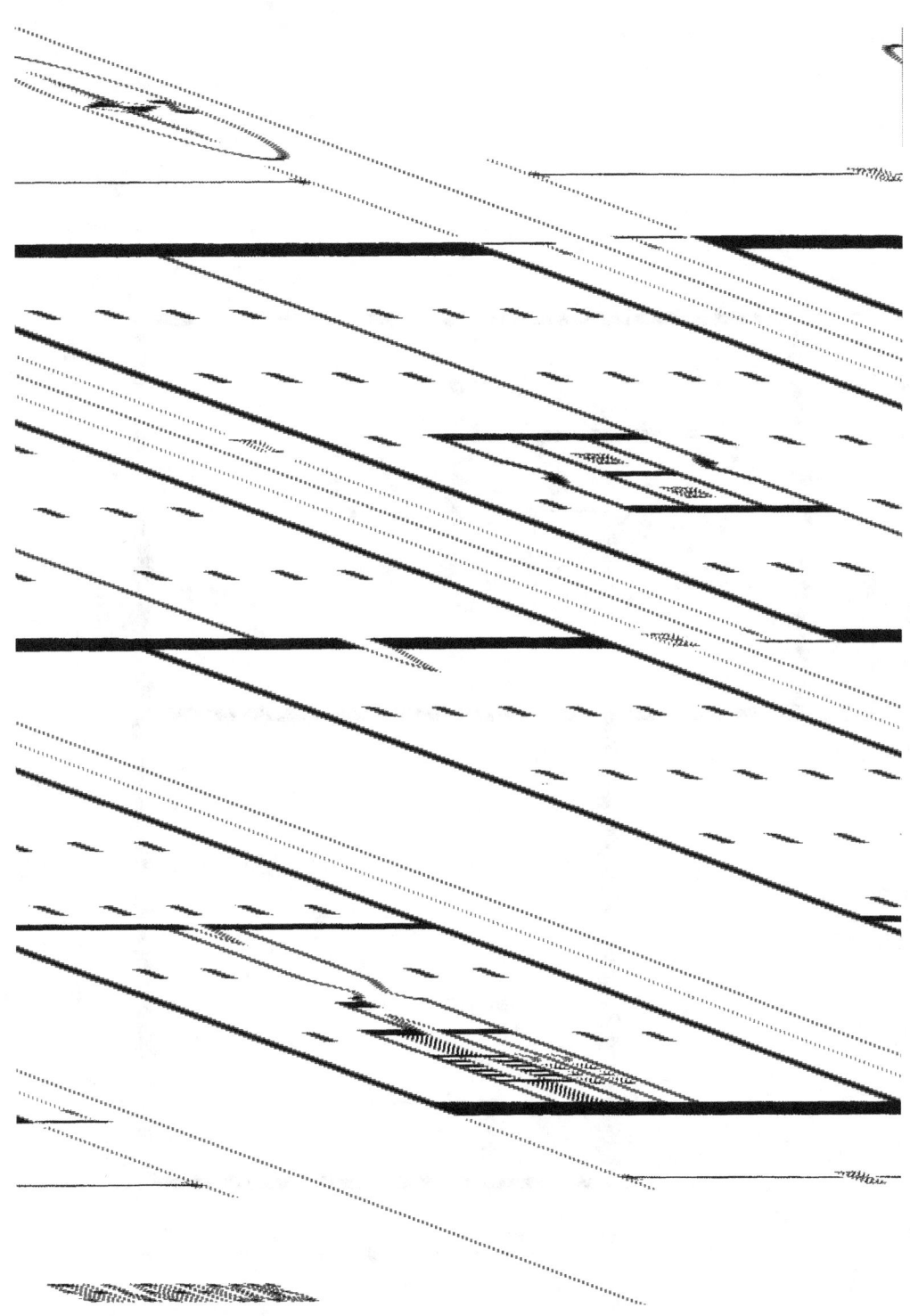

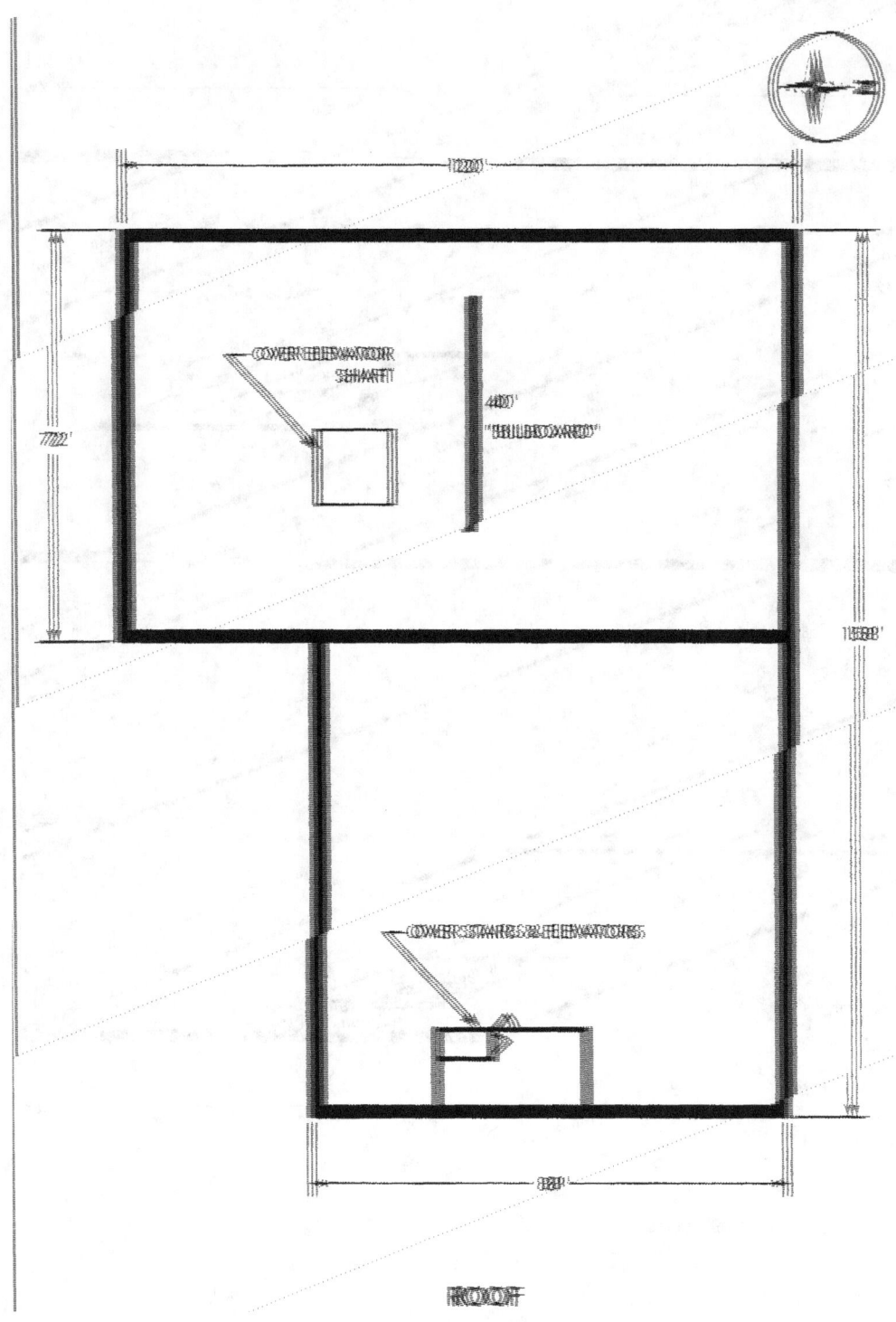

ROOF

APPENDIX E

Room of Origin

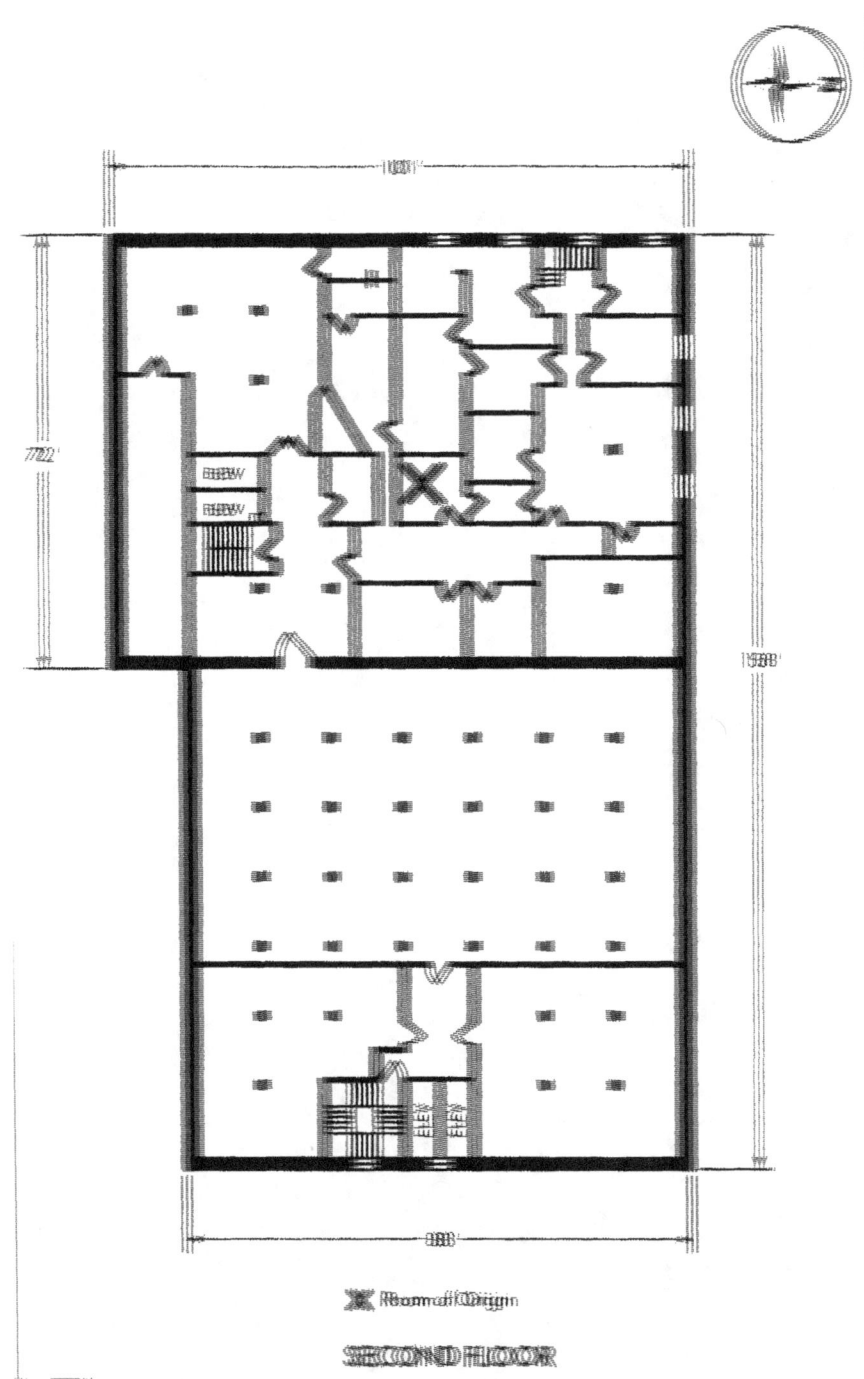

Room of Origin

SECOND FLOOR

APPENDIX F

Fifth Floor Fatality Locations

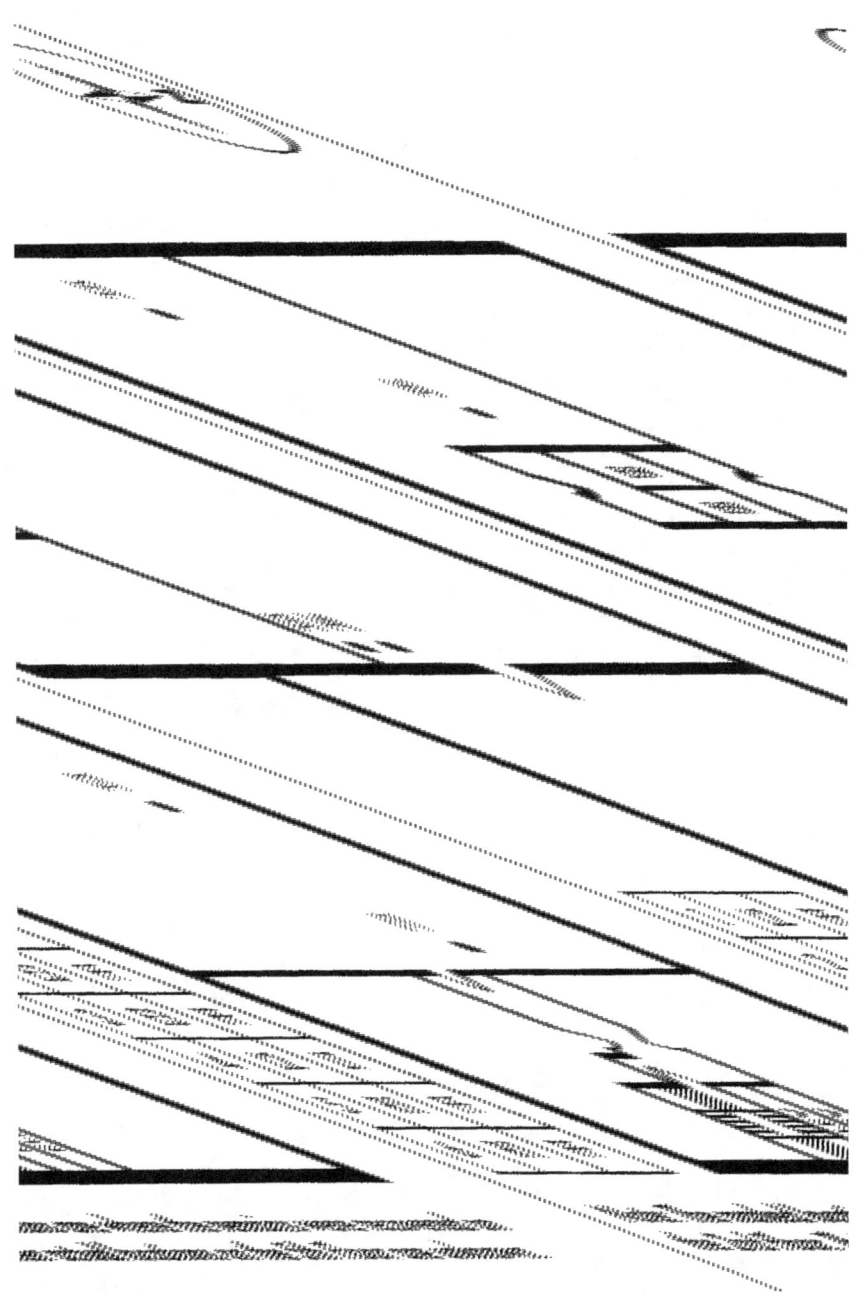

APPENDIX G

Timeline

A transcript of transmissions over "Operations A" channel was developed after an exhaustive review and many tape replays by the Worcester Fire Department's Board of Inquiry. This effort was critical in developing an accurate time line of events. Production of this transcript was made more difficult by the portable radio malfunctions during the fire. The following is a summary of the most significant transmissions:

18:13	Worcester Fire Alarm announces: "Striking Box 1438, Franklin and Arctic."
18:14:08	Engine 1: "heavy smoke showing."
18:17:57	Car 3: "…strike the second alarm. We are going to have the second alarm companies stage until we have a place for them."
18:20:27	Engine 13/Portable 2: "13 to Command, the fire is in the elevator shaft on the second and third floor."
18:20:46	Ladder1/Portable 2: "This is Ladder 100. I'm on the second floor and I'm in a freezer room, and I've got a room full of fire up here. I need a line on the second floor."
18:20:57	Rescue1/Portable 3: "… it's not Worcester Cold Storage. It's the building closest to the highway, we're up at the roof, we have heavy smoke and embers showing."
18:23:07	Ladder 1/Portable 2: "You are going to need that line up here to the second floor, chief."
18:23:10	Car 3: "They are on their way up, M. is going up there with Engine 13, they got a 2-1/2 coming up there."
18:23:22	C-300 (Car 3's aide): "Chief, I'm on the C/D corner of the building over here next to 13. We have heavy fire coming out the first floor, its not coming out yet, it's burning in and up."
18:24:01	Car 3: "…we have a 2-1/2 going up the inside now to the second and third floor."
18:24:08	Ladder 1/Portable 1: "We have a scuttle on the roof that's been completely vented now."
18:24:13	Car 3: "Okay do we have fire coming up through the roof yet?"
18:24:16	Ladder 1/Portable 1: "We have a lot of hot embers blowing through."
18:24:28	C-300: "…I'm right here with 13 and they are hitting it from the inside, that is where that 2-1/2 line is going from Engine 12." (1st Line)

18:24:45   Fire Alarm: "Command, be advised a citizen just reported to a police officer that there may be two people that live in that building."

18:24:58   C-300: "...charge Engine 12's line. Charge Engine 12's 2-1/2."

18:25:26   Engine 12/Portable 1: "...that 2-1/2 is off of Engine 1, not Engine 12."

18:26:04   Engine 13/Portable 2: "Charge that now, please." (2nd Line)

18:26:43   Ladder 1/Portable 2: "I'm on the second floor. Come up that flight of stairs and you go through a freezer chest, and I'm in the back where another freezer chest is. That's where you'll find it."

18:27:41   Engine 6/portable 1: "... there's two guys coming up here with a line now..." (3rd Line)

18:28:07   Car 2: "Command to Chief M." (Car 2 assumes Incident Command and Car 3 becomes Interior Command.)

18:28:13   Rescue 1/Portable 1: "Rescue to Ladder 100, you got one line coming up and you are going to have another 2-1/2 coming up very shortly."

18:29:06   Engine 12/Portable 1: "We are getting a 2-1/2 now, and we did find a staircase." (Stairway 2)

18:29:37   Engine 6/Portable 1: "Engine 1 charge Ladder 100's line."

18:30:08   C-300: "300 is at the elevator shaft on the first floor with Engine 12 and 13."; "They are hitting it with a 2-1/2 right now..."

18:30:45   Engine 1/Portable 1: "Engine 12, bring that 2-1/2 in here."

18:32:06   Car 3: "Rescue, did you check on that rumor that we have a couple of homeless people living at the rear of this building?"

18:32:13   Rescue 1/Portable 1: "Checked the second and third floor, we found nothing, Chief. We are moving our way up."

18:32:45   Engine 1/Portable 1: "Engine 1 to Engine 12 truck, charge that 2-1/2." (4th Line)

18:33:20   Car 2: "...do we have fire on the third floor?"

18:33:28   Rescue 1/Portable 1: "We have heavy smoke, no fire."

18:35:09   C-300: "300 to Command, we need a line on the third floor..."

18:36:29   Car 3 to Engine 2: "Come right in through the overhead door, we're going to go with 2-1/2 to the third floor so you are going to need a lot of line."

18:38:40   Rescue 1/Portable 1: "Chief, we are up on the fourth floor, we can hear fire crackling, but we can't see anything. We are in the rear of the building on the C side."

18:39:01   Car 3: "Okay, because it is running right up the shaft."

18:39:31   Car 3: "Put out an emergency broadcast to all companies operating inside to use extreme caution. There could be holes in the floor, and to use extreme caution as they are moving."

18:40:46	Car 2: "Give me a third alarm."
18:41:57	C-300: "We got the Rescue advancing a check on extension on the fifth floor. Now we're on the fifth."
18:42:09	Car 3: "Interior to Engine 2, just let us know what you got on the third floor when you get up there."
18:42:32	Car 3 to Scope 2: "…start taking these windows out on the side." (B side windows in stairway 3)
18:45:59	Engine 12/Portable 1: "Engine 12 to any firefighters in the second floor area, get out to the staircase, get out to the staircase."
18:46:12	Rescue 1/Portable 3: "Rescue 600 to Command. We need help, on the floor, below the top floor of the building, we're lost."
18:47:07	Rescue 1/Portable 3: "Rescue to Command. Rescue to Command, we need help on the fourth floor, one floor down, we're running out of air."
18:47:23	Car 3: "Last message, can you repeat? Last message?"
18:47:29	Engine 1/Portable 1: "Engine 1 to Command, get everybody out of the second floor, back them out."
18:47:44	Rescue 1/Portable 5: "Fire Alarm, Fire Alarm! Emergency, emergency! Clear the air, clear the air! Emergency!"
18:47:51	Fire Alarm sounds the Alert Tones
18:47:56	Engine 12/Portable 1: "Engine 12 to Command."
18:47:59	Rescue 1/Portable 3 (Portable 5's search partner): "I have an emergency."
18:48:01	Rescue 1/Portable 5: "Command, we are two floors down from the roof, this is the Rescue Company, come now, come now."
18:48:10	Car 3: "Where are you? Where are you?"
18:48:13	Rescue 1/Portable 5: "Two floors down from the roof."
18:48:17	Car 3: "Who is this?"
18:48:25	Car 3: "All companies we have an emergency. Somebody is two floors down from the roof."
18:48:30	Rescue 1/Portable 3: "Guys, we're not the top floor, one floor down."
18:48:39	Car 3: "What is your emergency?"
18:48:42	Fire Alarm: "Running out of air."
18:48:49	Engine 1/Portable 1: "Engine 1 to Command, get everybody out of the second floor, too."

18:49:20   Engine 1/Portable 1: "Engine 1 to Engine 12, ...back that line out of there! Get out of that second floor."

18:49:47   Rescue 1/Portable 1: "600 what's your location?"

18:49:50   Rescue 1/Portable 5: "Two floors down from the roof. Two floors down from the roof. Please hurry!"

18:50:29   Rescue 1/Portable 3: "We need air, we need air. We're sharing a tank right now, off of me."

18:50:38   Rescue 1/Portable 1: "P., if you need air, come on down. Come down."

18:50:43   Rescue 1/Portable 5: "We are lost, D. You got to send a rescue team up here for us."

18:50:52   Rescue 1/portable 1: "What floor, what floor?"

18:50:55   Rescue 1/Portable 5: "Second floor down from the roof, two floors down, I think."

18:51:04   Rescue 1/Portable 3: "We were on the roof, and then we checked the next floor down, now we are on the next one. Hurry."

18:51:39   Car 2: "What is your location? What floor are you on?"

18:51:44   Rescue 1/Portable 3: "Get up here! Please."

18:52:13   Car 2: "Fourth alarm, fourth alarm, notify Chief B."

18:52:29   Car 2: "I just struck a fourth alarm. What floor are those men missing, trapped on?"

18:52:36   Car 3: "They are on the second floor down from the top. Two floors down from the roof."

18:52:42   Car 3: "I have Ladder 2, Engine 3, and Ladder 1 going looking for them."

18:52:57   Rescue 1/Portable 5: "Fire Alarm we have a second emergency here. Get people up on this floor now or we are going die! We have no air, and we cannot breath."

18:53:13   Fire Alarm: "What floor are you on? What floor are you on?"

18:53:16   Rescue 1/Portable 5: "We don't know. We don't know. We were on a wall. We have no air. Please."

18:53:37   Rescue 1/Portable 3: "We are, you go up to the roof, you walk down the stairs, not the first floor from the roof, the second one down."

18:53:49   Car 3: "...activate your PASS system, activate your PASS system so we can hear you, activate your emergency alarm."

18:54:57   Engine 13/Portable 2: "1300 to Scope 1, drop one window down, and open that window up. You've got men inside."

18:55:23   Car 3: "...Activate your PASS emergency."

18:55:30   Rescue 1/Portable 3: "They are activated." (This is the final audible transmission from the Rescue 1 crew.)

18:55:32	Car 3: "Okay, Ladder 1, Engine 3, Ladder 2, they have activated their emergency alarms up there."
18:57:41	Engine 3/Portable 1: "Chief, we made it all the way to the top, and we hear no alarms on this side of the building."
19:03:35	Ladder 2/Portable 1: "Chief, are all people accounted for, out of the building? Ladder 2 and Engine 3 are on the fifth floor, we're still searching."
19:04:50	Ladder 2/Portable 2 to Portable 1: "What is your location on the fifth floor?"
19:04:53	Ladder 2/Portable 1: "Good question."
19:05:14	Ladder 2/Portable 1: "I believe that we are in the front part of the building."
19:08:28	Car 2: "Do you need relief for Engine 3 and Ladder 2?"
19:08:34	Car 3: "Engine 3 is already exiting."
19:08:53	Ladder 2/Portable 1: "Chief, get a company up the stairwell to the fifth floor. We can't locate the stairwell, or give us some sign as to which way to go. We are running low on air, and we want to get out of…"
19:12:06	Fire Investigation Unit: "Chief, we are going to have to get another line in the back. A 1-3/4, we have another fire in the ceiling here."
19:12:35	Ladder 2/Portable 1: "Send somebody up the stairwell to the fifth floor. Stand in the doorway and start singing."
19:12:45	Car 3: "Slow it down a little."
19:12:47	Ladder 2/Portable 1: "Get somebody up in the stairwell to the fifth floor. Have them stand in the opening and yell. We can't find the door back to the stairs."
19:12:59	Car 2: "Repeat the message, we can't understand you. Repeat the message clearly."
19:13:05	Engine 3/Portable 1: "Engine 3 has the message. Chief, we are going to the fifth floor, to the stairway, to lead them."
19:15:18	Car 2: "Command to all companies, close down operations, we want to give a yell in there, you see if we can hear from them. Close down operations."
19:15:56	Ladder 2/Portable 1: "Ladder 2 to Command we're done…" (This was the final transmission from the four member search team.)
19:16:17	Car 1: "Get the building owner down here immediately." (The Department Chief has arrived on the scene.)
19:17:37	Ladder 5/Portable 1: "We're on the roof at the stairwell."
19:17:42	Car 2: "Are you going to make any holes in the roof whatsoever?"
19:17:51	Ladder 5/Portable 1: "Negative. Chief, everything is coming out the stairwell and the elevator shaft."
19:18:09	Engine 3/Portable 1: "…I'm in the doorway on the fifth floor."

19:18:19   Ladder 4/Portable 1: "Command, they're calling Ladder 4. We are not...we are on the third floor."

19:19:13   Ladder 4/Portable 1: "It was Ladder 2, not Ladder 4."

19:20:14   Engine 15/Portable 1: "Change that 2-1/2 line for Engine 15..."

19:20:27   Car 2: "Command, all companies stay by the stairwell."

19:20:50   Car 4: "Deputy, be advised Millbury Fire is responding the scene, they do not have a radio, they are bringing a thermal imager, Okay."

19:21:40   Engine 16/Portable 1: "Command we have fire coming out the second floor in the front."

19:23:31   Car 1 to Car 2: "...how you doing with that line? We are kind of pushing back in on you. Do you want it backed off, do you want it backed off?"

19:23:41   Rescue 1/Portable 7: "Engine 12 charge that 2-1/2..."

19:26:57   Car 1: "Give me a fifth alarm on this right now, a fifth alarm."

19:28:45   Fire Alarm: "Car 1, it's going to be Engine 5, Engine 9, and Ladder 6 are your fifth alarm companies."

19:28:53   Car 1: "10-4, tell them they are to be the RIT Team, and they are to assemble and meet up with Car 200 ...on the left side of the building."

19:31:47   Car 1: "...I've got a thermal imager down here from Millbury, and I want to send it in. I am bringing down an aerial scope on this side of the building, on the east side of the building. And we want to give a couple of guys down here to take in with them."

19:32:07   Car 3: "Okay, nobody in without lifelines, though. We want lifelines on everybody. We have ropes tied off upstairs."

19:35:54   Engine 7/Portable 1: "We need some guys in the stairway on B side off the loading dock, there are two 2-1/2's up there, and a lot of fire, and three men up there."

19:36:06   Safety Officer: "...All fourth alarm off duty men meet behind the Rescue."

19:37:17   Safety Officer: "All companies, REHAB is behind Ladder 5 in front of the building, REHAB is behind Ladder 5."

19:38:32   Car 3: "Ladder 2, Lt. S."

19:38:39   Car 3: "What's your location?"

19:40:35   Car 1: "...I don't want any individuals going in this building unless they are teamed up with two or three people."

19:43:06   Car 1: "...I have Engine 8 going into operation on the diner side of the building. The D side of the building."

19:44:10   Rescue 1/Portable 7: "We are never going to make the fifth floor from the stairway."

19:44:13   Car 3: "Okay, don't risk it. Don't risk it, back down, Back down."

19:44:20   Rescue 1/Portable 7: "...we can only make four floors."

19:45:00   Scope 2/Portable 2: "The Millbury guys went in by the stairway. They could not make it at by the window."

19:48:04   Engine 9/Portable 2: "Engine 8 Group 4 the structural integrity of the building has been compromised. Keep everybody away from the building."

19:48:17   Engine 9/Portable 2: "The 290 side of the building is in danger of collapse."

19:49:08   Car 4: "...where is Millbury with that thermal imager...We are on the back stairs here....It is very hot on the third floor, we're trying to make headway, we could really use..."

19:49:47   Engine 8/Portable 2: "We are coming up with the thermal imager now."

19:51:04   Fire Investigations: "...the fire just vented through the roof on the C side. Heavy, heavy fire vented through the roof."

19:53:12   Ladder 1/Portable 1: "There is a crack in the wall. Right over the ladder in front on Ladder 4. At the top floor, it is about a 6 foot crack."

19:53:50   Car 4: "Chief, be advised the imager has stopped working."

19:54:02   Car 4: "We can not make it to that floor. There is to much heat...we are going to need lines up there again."

19:56:14   Car 3: "We're backing out the back. We got a report that the walls were weakening in the front. We are trying to back the lines out, so we can use them. It is through the roof in the back, and its going like hell right up the side. I think we are almost ready to go to an exterior attack."

19:58:10   Car 3: "Command to all Companies evacuate the building...sound the evacuation signal...evacuate the building."

www.ingramcontent.com/pod-product-compliance
Lightning Source LLC
Chambersburg PA
CBHW081228170526
45165CB00009B/3003